Carbon 14

Ann Deagon Carbon 14

University of Massachusetts Press, Amherst

Library of Congress Catalog Card Number 74-79483
ISBN 0-87023-170-7 (cloth)
ISBN 0-87023-171-5 (paper)
Printed in the United States of America

Library of Congress Cataloging in Publication Data

Deagon, Ann, 1930-
Carbon 14.
Poems.
I. Title.
PS3554.E116C3 811'.5'4 74-19348
ISBN 0-87023-170-7
ISBN 0-87023-171-5 (pbk.)

"Poet in Residence" is reprinted by permission from *A.A.U.P. Bulletin* 58, no.4 (Dec. 1972). "Researches" is reprinted by permission from *A.A.U.P. Bulletin* 60, no.2 (June 1974).

"Arcadia," "The Last Eureka," and "Samothrace are reprinted by permission from *Beloit Poetry Journal* 21, no. 4 (Summer 1971). "The Sibyl's Bath" is reprinted by permission from *Beloit Poetry Journal* 23, no. 2 (Winter 1972-73)

"The Wanderings of Penelope" is reprinted by permission from *The Carleton Miscellany* 13, no. 2 (Spring/Summer 1973), copyright by Carleton College, August 23, 1973.

"Dependence" and "The Death of Phidias" first appeared in the *Carolina Quarterly* 23, no. 2 (Spring 1971). Reprinted by permission.

"Afternoon of a Tourist, Part I" is reprinted by permission from *Coraddi* (Fall 1972). Copyright: Fall 1972, *Coraddi* UNC-G.

"Final Strophe" is reprinted by permission from *Crucible* 8, no.2 (Spring 1972). "Niobe at Moses Cone" is reprinted by permission from *Crucible* 9, no. 1 (Fall 1972). "Itinerary" is reprinted by permission from *Crucible* 9, no.2 (Spring 1973). "Four Stones" is reprinted by permission from *Crucible* 10, no. 1 (Fall 1973).

"Afternoon of a Tourist, Part II" is reprinted by permission from *Epos* 23, no. 3 (Spring 1972).

"Invoke the Muse" is reprinted by permission from *The Greensboro Review* 15 (Winter 1973-74).

"Daphne on Woodbrook Drive" is reprinted from *Miscellany* 7, no. 1 (Winter 1972); "Le Figlie di Dracula" from *Miscellany* 8, no. 1 (Winter 1973); "Nude Aging" from *Miscellany* 9, no. 1 (Winter 1974). By permission of *The Miscellany, A Davidson Review*. Copyright 1972, 1973, 1974.

"Roommates" is reprinted by permission from *Poem* 14 (March 1972). "The Sibyls in the Second Class" is reprinted by permission from *Poem* 21 (July 1974).

"Niobe at Moses Cone" is reprinted by permission from *Red Clay Reader* 7 (Nov. 1970).

"Nausée" is reprinted by permission from *St. Andrews Review* 2, no. 2 (Spring/Summer 1973).

"Carbon 14" and "Icarus by Night" have appeared in *Smart* 1, no. 1 (June 1, 1974). Reprinted by permission.

"Reflection" is reprinted by permission from *Southern Humanities Review* 6, no. 2 (Spring 1972).

"Once upon a Greek" is reprinted by permission from *Western Humanities Review* 26, no.1(Winter 1972). Copyright by *The Western Humanities Review,* 1972.

FOR DONALD

Contents

Invoke the Muse

I string my kite for storms
with picture wire
wind it round my fist
clench teeth and hold hard.
That singing strike
that melts your fillings
that's inspiration.

I had a friend once
walked by the beach
and drew the lightning.
It fused the zipper to his crotch.
Fool coroner!
If they'd unzipped him
there was a poem inside.

Arcadia

Back in the war, when meat was rationed
we kept rabbits out in the backyard
far enough away so we wouldn't smell them
but still we treated them about like pets
except for eating them.

We killed them clean, no blood or fuss
a sharp blow to the skull or sometimes two.
Cleaning them of course was a little messy
but still the skins were white and very soft.
I made some into muffs.

That's why it's hard to figure out
what that dream was all about
or what white rabbits had to do
with that black man that looked like you.

The way the dream went, William, was like this:
a bunch of friends of mine were at a farmhouse
out in the country, and we heard this noise
a sort of random thumping from the yard.
They asked me what it was.
I told them rabbits
thumping the field mice
standing on one foot, flattening them with the other–
they did it all the time.

The people said they'd like to see that thing
and so we all walked out into the half-light
watching the rabbits as they thumped around
and pretty soon we started out ourselves
and thumped some on our own.
And then we started
thumping the rabbits
standing on one foot, flattening them with the other–
it was a lot of fun.

By now our feet were flat like skis
from crotch to toe enormous Z's
bony powerful and spare
pelted with white rabbit's hair
and every buck I'd smash with mine
I'd feel the shock go up my spine.

It finally woke me up. I lay there wondering
but soon I drifted off to sleep again
seeing that party at your place last Easter
hearing the throb of music and the laughter
the beat that I couldn't dance to.

The people finally left, and from the door
I turned to see him coming down the stairs
holding a wooden flute, and halfway down
he stopped and looked across at me and raised
the flute to his lips and played on it.

Oh William
was it you?
He looked like Pan
I was sure his thighs were rough with shaggy hair
some sort of jungle stalked me from that stair . . .
God, William
it it true
dreams make the man?

Dependence

There is something I have been hiding
all these years.
At six I let it loop between my legs
to play cowboy (you like that, Freud?)
my lariat.

At sixteen I wound it up my eve
to nip my apples
my secret snake.

At twenty I took my razor out
and slashed it off my navel
stepping bloody
to your bed.

Now forty I admit
it has grown back again. I wear it
in coils around my hips. At night unwinding
it slithers over you asleep
and nuzzles for its hole.

Sleep, love, until I find
a sharper razor.

Niobe at Moses Cone

I

I must have been inserted through some slot
to lie this white and flat
in my sheet envelope
in this deadletter office of a ninth-floor room.
God knows what's written on my face
except thank God there'll be no baby this go-round.

Before I met him, when the nights were hot
I'd sleep spread out like that
and there were times I woke
and felt a strange hand probing toward my empty womb.
But I knew better than to trace
that hand to my own shoulder—so I made no sound.

> This afternoon again my hands are strangers
> fumbling the alien stiffness of the sheet
> fluttering semaphores of unseen dangers—
> as distant now as Alice's long feet.

I met a lady once at twenty-four.
She was a dowager of fifty-five
who'd lost her mother recently, and I
dear God, reminded her of darling Mother.

She had to be the little girl once more,
it was the only way she could survive.
It's comic how indecently we try
to hide our dread of dying from each other.

> We lie all children in these beehive wombs
> waiting to be reborn. We do not weep.
> Silent we watch starched hope patrol the rooms.
> Only the T.V.'s sob themselves to sleep.

II

Last spring we lived together out from town
a farmhouse with tall windows—and I wrote
"Bright and birdly wakes me up
day at the eaves nesting."
That was just last May.

In June I woke up to another sound
a cackling screech that tore at my own throat
stumbling wrenched the window up—
there on the ledge resting
the spring-coiled cat lay.

A half-grown guinea chick that fell
somehow into the window well
flapping, squawking for dear life
brandished his terror like a knife.

I screamed, the cat leaped over and was gone.
We took the guinea to the neighbor's place.
I think we knew
it was an omen, and the house a snare
and all that birds of morning sang was fear.

Late afternoon we picnicked on the lawn.
I only saw it happen in his face—
a child we knew
the truck so fast she didn't have a prayer
the gravel, bloody screaming in my ear . . .

We left the next week, moved back to the city.
Whatever came between us, guilt or pity
we saw the thing was over, parted friends.
Why should we last, when everything else ends?

III

Night comes premature—a claw of light
strikes at the aerials and tears
a stillborn darkness out of the sterile clouds.
Again October with its hungry winds
tonguing the rough bricks
panting for blood licks
these cubicles where every pretense ends
these well-provisioned cells, these larval shrouds
scenting the honey of our fears.
My October snarling prowls the night.

If God would throw
the window up and scream away
this nightmare lion's den
if I could know
a morning ends the dream someday—
I'd loft my life again
surge to live and rage to die
under the Abyssinian eye.

Roommates

Opening up the cottage
at the end of summer
we smelled the squirrel
fallen down the chimney
trapped
when the damper sprang back
and saw the wood between the window panes
around the casement
gnawed in mad rococo
scalloped with desperate elaboration
        she made her mark, that squirrel
        she found
        something to sink her teeth into.

Meeting you sometimes now
I scent your loneliness
I see
around your eyes
something has been gnawing
desperate for life
        for god's sake don't smile
        I know your teeth
        are worn to the gums.

Daphne on Woodbrook Drive

Brushing my hair at a window into spring
across the budding woodlot I can see
the blasted treetrunk where once lightning knifed
a palate of smooth fleshtoned wood now weathered grey.

Winter and summer I watched an image form
a human face emerging from the trunk
the ragged bark encircling it like hair
its features still distorted by the pang of birth.

Today I walked across to see it close.
Deep eye thin nose and twisted mouth dissolved
into a labyrinth of hollow trails
where some unseen untiring minister devours
his destiny.

Elaborate the minute passageways
construct destruction build their world's collapse
yet down the intersecting corridors
minuscule clustered eggs project the race into
infinity.

Brushing my hair at this mirror into time
I see the tracks of countless smiles and sorrows
wrinkle my image honeycomb my face.

My hair has weathered grey and the patterned bark
encircling first my throat benumbs my senses
spreading around my mouth around my eyes.

Is this Prometheus' secret that I should
see now my face imprisoned in the wood
caught in a myth I had not understood?

Nude Aging

Whenever I run the tub
one silverfish or two drift in
that porcelain ocean
where I founder
rotted with memories.

Crossing to Crete
on the Ariadne
by moon the dolphins leaped
in a delirium of silver spray.
So sound and surface
in that deep chased design
my tearing dreams,
and with that same
high marble eye I watch
my lifeline swell and wrinkle,
soak in the drowning present
knowing
another ocean will turn me
wrinkled without the water
turn me
silver without the moon.

Night Rider

How wound in these quite ordinary sheets
in this flat oblong bed in this
tall oblong room myself at forty
imperceptibly stiffening with rigor vitae
continually encounter that galvanic dream
air borne skimming a virid forest
gliding tree by tree an infinite arboretum
grasping springy branches catapulted
soaring in a green sunlight clasping
between my thighs resilient treetops
ethereal canter jolting me out of sleep

Jung Eliade Campbell my masters
riddle me this nightmare of dear riding
where my bursitis pricks awake
the evolutionary ape
still vaulting arm by arm
the leafy interstices of my mid-brain
or do I mount bareback and ride
the fantasy of still pubescent Greece
some curly centaur tangling my hands
in that wild tawny chevelure
maning his sculptured back
and wake disheveled
my fingers
pungent with resin and love

Reflection

Wrong, Plato, wrong!
This mirrored bed–
once carpentered and twice reflect
in haggard eyes
that hawk the real–
out-beds idea.

But Diotima knew
we slake in effigy our wanton's want,
writhe the flesh out of tangled time to glimpse
the mirrored moment's bright epiphany:
Silenus
whose grapes cluster
the prancing thigh.

Afternoon of a Tourist

I

Down the Autostrada del Sole
lofted on immemorial arches
of forgotten aqueducts
channeled through the gallerias
we flow into Rome
swirl through the piazzas
gurgle at the fountains
drain into the illuminated basins
of decaying hotels
(my cousin threw up in the bidet
at the Plaza Hotel in 1967
we had diarrhea at the Margutta)
shall we visit the cloaca maxima

What did we come here for
jostling in the ruins of dead fora
viewing museums
                                        Love, love
press me against this ancient stone
like that one pear tree in Haute Savoie
espaliered against a crumbling wall
heavy with fruit

II

The streets of Pompeii
are crowded with the living
strolling along our Kodacolor, striding
unforeseen into our Ektachrome.
We students of the past
prefer the ones immortalized in plaster
encased in the museum. We are enamored
of the aesthetics of the waspsnest
after the wasps are gone. Humanity stings.

Unearthing
my buried cities, retracing
what roads converge on this hotel
I sketch al fresco the mythic union
of bulls and maidens, Bacchus
sleeping, his mouth ajar.
I have moved unmoved among these people
untouched, untouchable, my childhood
burning in recollection, our long encounter
resonant in my ears, my eyes blurred
with memory.

We have lived too much inside, let our volcanoes
tomb us with ashes. Walk with me, love
along this waterfront scaling with refuse
of food and drink and excrement, see these faces
dark with the sun, grip these hands
brutal with life.

In Camera Obscura

Too often lately
my eyes have strayed
across the beveled mirror that reflects
our bedding
to the prismatic edge
where an un-world of color, clashing in planes
kaleidoscopes into a rage of light
the subdued furnishings of our ten years.

What message shall I leave
walking out of our mirror, out of your life?
I do not have illusions. I know
now here is forever framed in nowhere,
movement is an illusion of stopped frames
and all loves out of focus.
We have seen each other through a glass darkly;
I do not expect to see anyone more clear.

I leave this dark room, love,
only to walk a little in the sun
wear green while my season lasts
and then take on
the kaleidoscopic colors of decay.

I
PENELOPE AT HIGH RISE

My floor is wall to wall my walls
are floor to ceiling these flat windows
hermetically sealed against the gangrene air
GERM-FREE ENVIRONMENTAL LIVING cleanliness
is seventeen flights up and god
must run the elevator. I have felt
his presence solitary to the wink
of lighted buttons in the lift and sink
of my miraculously unembalmed yet
viscera.

So we descend to Naples the jet braking
pulled at my gut and one thought twitching
sparrow at worm: who was it sang
I will show you how the lilies blow
on the banks of Italy? Sweltering city
streets filthy men filthily brash
at Meta Felix Tours a girl confirming
our reservations tilted her head reached up
to pinion a dark wing of hair escaping
as on some Bacchic vase head thrown back
arm ripe as thighs bent to the air and through
the flimsy blouse I saw her armpit tufted
with black—as though a mist of spray
some dark hung with seaweed cavern where
Odysseus malingered.

My life is wall to wall dreams only
seep in from that uncleanliness outside
to weave their trails across my nights and days:
some foetal swelling under the carpet mounds
springs to a sapling thickens into trunk
bursting the roof with branches a tremor
rising through all the stories shakes the bole
peaks at its blossoming crest, drenches
this room in lilies.
In such state I lie.

II
PENELOPE AL MUSEO NAZIONALE

Dimly through the glass-paneled door
the dancing faun in the Hall of Great Bronzes
CLOSED FOR LACK OF PERSONNEL the guard
insistent at my arm "Originale,
signora" lured me reluctant
down the vast grimy CLOSED FOR REDECORATION
corridor, behind the temporary screen
encountering a marble boy, near man and
near to life size, new unearthed, unearthly,
clay hardened to his tilted throat,
his armpit, where the turning thigh
curled into the mutilated part.

Leering nudged me toward a packing crate
tumbled with fragments: toes, fingers, feet,
a nose, a clump of hair, god knows what
debris of crumpled statuary—lifted
clear of the jumble one small perfect hand
fingers half curved as if to cradle flesh,
gestured toward my handbag, spoke in English
"No one will see." I shuddered, thrust away,
fled down the watching corridors, returned
to the hotel room.

Rinsing out my things
I stand before the mirror wondering
how all these toes and fingers and soft hairs
these fragments of my body so smooth joined
have come to rest at random in this crate
CHIUSO PER MANCANZA DI PERSONALE
and no Pygmalion.

III
PENELOPE AMONG THE RUINS

Pompeii well frequented whore on Sunday
displays her inner chambers unabashed
to our post-doctoral examination.
Can you remember before Vesuvius
laid you steaming pillowed on lapilli
before the kings of Naples
rifled your bowels with rough tunneling
before you went public? What was it like
your girlhood—hot like mine
in Alabama always the sun staring
like the fable hot to make us shed
our damp dresses languid on day beds
drawn shades yellow with frustrate light
to nap in apprehensions of
dark bodied strangers.

So our volcanoes our dancing fauns
internal to imagination sickened
into a leprous ladyhood that smiles
to munch its crumbling fingers with its tea
and views the ruins unshaken.

IV

PENELOPE UNRAVELED

Is poetry the cure or the disease?
Across this printed grid I chart my findings
digging up the past dredged up my own remains.
(At Stabiae with brush and pincers joining
flecks of plaster other craftsmen fledge
that breathing skim of painted fantasy
that melts a room to myth.) This mural, friends,
these bits and pieces of mosaic days
portray Penelope by stealth unwinding
herself, myself, ourselves: you write a poem
meant to be mythical, you strike a pose
somewhat remote, construct a cardboard figure—
some bloody condensation of the air
seeps to its veins, you slash her
and I bleed.

Orphée

Your eyes close like gates, like
the lids of coffins, my kisses
weight them like coins.
You let me play your body
like a harp
but no way opens
into your dark.
Behind your lashes' shimmer
your dead youth
embraces your youth's bride.

Go, singer
sing her those songs
whose lilt and fall
unbound my mind.
Where there is no choice
I choose the light
though there are days
when sunlight
sears the eyes
deeper than hell.

Dagon

I

This dead shark
luminescent to the archetypal moon
stilled sperm at glimmer
        when was the last time
        we lay in sand
        forgetting
        that scythe archaic smile
        that sculptured eye
        blind as marble
        Aphrodite
        rising from the sea

II

Today at Sperlonga
in the gallery
the fragmentary anguish of Odysseus
lashed to the helm
thrust through two thousand years up out of
whirlpool earth
        we have made love
        love has made us
        time swirls
        in the shark's eye

III

Sinking
through this dark riven water
writhe toward me love
waist downward
scaled with my kisses
        who knows but this tide draws
        to some deep grotto
        where sea lions slumber
        in a shimmer of light

Erotic Fragment

When my mouth
is wet with you
and I have dived
into the deepest grotto
of my old fantasies
and arching up
from the green bursting depths
beach in your arms . . .
always there
the stone is shattered
the papyrus rent
and nowhere
any conclusion

Itinerary

I
ARCHAEOLOGICAL JOURNAL

These hard hollow rolls of morning in Rome
like all rinds of cistern, bath, mausoleum
shells of temples with god scooped out,
that lumbering voluptuous tortoise antiquity
dry hull and thin soup dribbling
down Italian chins.

We jostle
in empty forums, along the fleshless
corridors of museums where none of all
myth-lapped vases brims with milk, with wine,
not one dead child in honey long embalmed.
The slow lecherous tongue of time
has sucked their flesh away. Clodia,
Clodius, your lovers have devoured you,
left me no morsel.

In the streets
of Rome at evening stroll your only relics:
from Virgin's womb, suckled by wolves and leopards,
insolent, ithyphallic—while I lie
hollow with longing in this hollow bed.

II

AN ORACLE AT DELPHI

Six birds at Delphi cleft the ring
of my viewfinder, spun me with their swirl,
settled on the ruined god's high Doric
capital in a flurry of light.
Here poised the world at center, balance point
between the suck of earth and soar of air.
All tension is oracular. Wherever
air caves itself in earth or marble thrusts
mountain or column to the upward chasm
space springs resonant with the twang of time.

Omen me, Greece. I am battered with your stones.
I have climbed your broken stairways, passed
between the tumbled courses of your marble cells
embraced your columns, penetrated nude
into the brutal clash of molecules
dense whip of atoms goaded by your sun.
The crush of time against my flesh inflames
a passion of want. Where shall I find you whole,
land of the broken phallus, mutilated herm ?

III
OLD MAN AT THE DANCING

Why did it surprise me—
baggy trouser legs behind the scraggly low
tree watching the path to the theatre
nothing else visible, no movement, only
the jerking hand. Was it Apollo
embraced that way a girl gone all to leaves?

Old man who saw me blurred in branches young
enough to swell your crippled fantasy,
I hurried on to watch the dark young dancers
leap to the whirr and twang of ancient modes,
myself voyeuse. One dance is oldest
and I no less its dancer, you no less.
Odysseus waking in the thicket watched
swirling maidens play at ball, strode forth
holding a branch athwart his thigh to shadow
the clench of beauty at his loins.

Once upon a Greek

Old Aristotle omne animal
you say post coitum is triste
I say Ha!
You Greeks would try a woman once
you'd know
better.

The Sibyls in the Second Class

Il conduttore leans
hands behind his back against the glass
partition of our compartment. We museum
red palm white fingers nailless but one
(left fifth—grey pearl of snot and earwax)
whorls grimed intricately rococo
(he has stamped them on our passports) one scar
black at center where some lurch
skewered a pencil, the welling blood
long dry to memory but reddening
in the blunt wide palm. Our nostrils flare.

After the bronze Etruscan handle-hands
marble Greek fingers, the muted stir
of mummies' hands clutching at life,
this live exhibit (ante mortem) past
scarred into lifeline, pulse at rhythmic swell
incants us to prophetic trance. We hover
tracing an eager palmistry across
our still emboweled victim—while he stares
down the Campanian orchards, our reflections
palmed beyond their glass encasement gesture,
flitting Vergilian at the edge of sight,
waken his restless vigor and confirm him
a man of destiny.

Le Figlie di Dracula

The good Catholic girls of Rome in black
shoes, hose sheer but severe, black bottomed,
breasted, age into motherhood, conceive
immaculately still in black—or haunt
in habit the basilicas, sweep the trodden saints
with their black hems, within the cloisters
rehearse the hour when gathering the dark
folds of their martyred flesh about them, whole
they lie down holy in the catacombs
waiting the sun to set. Their teeth grow long
anticipating resurrection, nostrils
flair the incense of burning flesh, their eyes
unblink the agate stare
of all saints.

Monuments funéraires

I

Philocles
set on his daughter's tomb
her statue, marble.
Now remain
only ten toes
two sandal thongs
parting them lightly
and a space
that lilts.

II

His sword
they reddened in the pyre
to hoop the jar of ash
around whose girth
the painter slung
the prance and blazon
of the field
and set upon his twenty years
the colophon of art.

III

Two years a child
three thousand years a clutch
of cradled bones
inside the honied jar.
Forever now
on view behind the glass.
Is it sweet,
immortality?

Four Stones

Two friends gave me four stones from Pompeii.
They spoke what stones speak, man-shaped
the crust of mortar crackling
having escaped out of the patterned hall
the terrace
into the infinite mosaic
of now.

They tumble in my stamp box, rasping
the pale paper
whose roots moulder
high in some stretch of western forest
and nourish scarabs.

The Sibyl's Bath

When the old sibyl began to spit blood
three of us climbed eastward through the gorge
along the rough terraced vineyards seeking a girl
grown up in sunlight. We saw her first
dancing alone between the olive trees
a young goat capering at her heels, her hair
aflame with light. I felt the sure breath of god
prickling my scalp. Her people consented.

After the old one died it fell to me
to lead the girl down through the caverns to the bath.
She understood we were in the womb of earth
but somehow she was not prepared to meet
divinity. When the lamp revealed him
rising soundless through the black water
gnarled, ancient, white with absence of the sun
larger than common frogs but yet no giant
she began to scream, was screaming still
when I covered the lamp and the water closed around her.
She had not conceived of the god in such a light.

I carried her to her couch in the inner cavern
and soothed her babbling, dried her, gowned her fresh
in the sibyl's robes. When I saw
her hair had turned white I knew the god had chosen
well. Whenever ambassadors came
I had only to meet her eyes to see the shudder
rise in their depths, the holy shiver seize her
and that sweet barbarous hair-raising lilt and fall
of unknown tongue. The god was well served.

She began to visit the bath more often, sing
to the god, lure him to rise, trailing
her long hair in the transparent blackness.
Who could foresee if not the god himself?
We had no sign until that day the caverns
welled up in uninterpretable sound.

She stood beside the pool, the writhing god
dangling before her: each hand gripped one leg
and with a blinding shriek she flung her arms
wide above her head, her upturned face
and hair festooned with blood. And still from her throat
that deep rough croaking outcry spurting.

Now she is calm. We have told the ambassadors
to return at autumn. We must decide
whether or not to go on. If we could know
the meaning of the act: was it his will
to die, did he possess the virgin, goad her
to his dismemberment—or could a girl
conceive and nurse in such translucent gaze
a purpose as demonic as a god's?
We wait for signs. She watches blankly
the croaking laughter dribbling from her mouth.

Position Paper

Under the Colossus looking up
I tell you Freud it isn't envy
that gets me down. The inscrutable scrotum
is just a bag of marbles played for keeps
and he can't move the world for all his lever
unless I let him stand on me.
So I ask myself what am I doing here
under the Colossus looking up
musing on that branch well hung with cherries
that bowed before the queen of Galilee.

The Death of Phidias

Between the trial for embezzlement and the trial for impiety
Phidias sickened in prison and then went mad.
When we brought his water he flung it on the floor
and scraped up the hard-packed clay with his rotting nails
to mold crazed figurines:
a man with his head attached between his legs
and on his shoulders a great erection,
women with holes in their breasts and teats on their buttocks,
babies with too many arms and not enough legs,
a hunched hermaphrodite with a giant hand
coming out of its rump like a rooster's tail.

When they put him on trial he crowed like a rooster himself
and when they asked what he meant by that he said
he was Zeus the Cock crowing so the sun would rise.
They convicted him, but some of the jurymen wept
and all of them shuddered. Back in prison
while his friends were scraping up his fine
he ate the crusts of his bread but molded the insides
with his saliva into indefinable forms—
intestines that flowered into cabbages,
livers with claws, things without names or existence
except in his hands and our half-tainted eyes.

He began to save his excrement in a corner
saying that it was his earnings to pay his fine.
That last day when we found him he had torn
one wrist with his toenail, blending the oozing blood
into the lumpy mass. It lay beside him,
his masterpiece self-portrait, like him dead,
only a little more stinking than his flesh
and not much difference for long between them.

We buried it beside him, never spoke of it.
We jailers learn too much we don't dare tell.
Some nights I dream that the whole acropolis
quakes into chaos and the long walls crumble
golden Athena melts and this bright air
glooms into prison dimness and the stench
of Athens rotting.

Nausée

Must we now
up from the gurgling sink-holes of nightmare
grope the sodden handfuls of our growth
plaster the mud-child with the dough
of tumored breast and hip
the yeast frothing?

Phidias, Phidias
make us to marble
if only to feel
the clean, the pitiless
chisel.

The Amazon at the Metropolitan Converses of the Past; the Female Poet Replies

Understand, Companion, we were not
men unpenised. (Our sex inheres:
the blade that pierces it cuts off our lives.
We die as women still.) Nor did we mask
our bodies into maleness. That black-figured vase—
see how the painter burnt our single breasts
into the reddened air, what hips bestrode
those backboned horses. We were women, Friend,
but with a difference. To draw the bow
full to the ear and loose the singing shaft
requires a harsh economy of form.
We shaped us to our art. The knife's poiesis
created us half nurse, half warrior.
(We know it cannot be passed on. Our daughters
bud relentlessly into the time of knives,
the time of choice.) Is it still so with you?

> Huntress, would you have it change? The choice
> still makes the chooser. We who hunt the word,
> who nurse at breast that sharp malignancy the muse
> still slash our bodies into music, still
> enact the halving of the moon upon
> our womanhood, bestride the nightmare still,
> still from self-sculptured breast we spurt
> the milk of love, the blood of art.

Summer Nights in the Dormitory

(In Plato's fall we've fallen all
from eidos down to eiderdown . . .)

Some mad Sade clenching his teeth in my pillow
snaggled a drift of stifling fluff
smothered my face in a flurry of doves
skimming fingers tongues eyelashes.
Gasping awake catching at floating dreams
I downed the feathers, bred them to my body
glued in my night sweat
along my long neck, my collarbone
between my breasts adrift in down
down where the hipbone gaunts
from the hollow belly, down
where the thighs crease hotly.

Watching from the other bed
gowned in her archaic smile
she murmured as I pinioned her:
Me Leda, you Swann?

Researches

Scudding down the icing roads
in a snowstorm Leif and I
discovered America the gas station
crackling with lightbulbs St. Esso's fire
you lay on your back in the snow putting on the chains
a Norseman licked toothsome out of ice
blond hair aglint with snow

While our fingers stiffen wrestling chains
in the rout of Marathon
a Persian ship in panic puts to sea
Cynegirus' hand gripping the stern
braceleted with a Persian axe
above the plain an unidentified shield
blinking the Attic sunlight signals treachery

O constant scholars
for whom still glints
that unknown shield across
two thousand and a half such revolutions
as whirl us out of springtime into snow
whose Grecian eyes still flash
a sunlit universe
to universities of inclosed rooms

Beyond these snow clouds Leif
far fixed stars long cold by Marathon
throb their wounded light
continually as scholarship
more faithful even
than your chill hand between my thighs
than your hot tongue between my teeth

Sphinx Ludens

Thebes of the 7 gates in
to whose maw
generation up
on generation
all potent heroes come
who tread the earth

and at whose out
skirts shrivel the de
tumescent un
entombed fore
fathers of the epi
goni

                im
penetrable city
spread your gates
we two come in to
lay us down with Laius
I your casta
you my swollen son in
cestuous and zestful re
enact
the play of Thebes

The Last Eureka

I have deciphered Linear A.

In my wine-dark unfathomable bathtub
or while besquatting like a colossus
my everflowing cloaca maxima
or brushing from between my cavities
the relics of Odysseus' crew
I have meditated masturbated mastered
the Secret of the Past.

Epigraphy paleography graffiti
scarred stone, the lacerated clay
skins flayed from unborn lambs
crushed reeds encode
a mummied past that begs our question.
All inscription
undeciphered mocks our humanism
deciphered mocks it more.
(Sappho in strips
wrapped a grinning mummy grinning still.)

I have perfected my transliteration.

On transparent squares of pine pulp dabbled
red yellow brown earth colors to earth men
I now bequeath the ultimate translation
of those ten thousand tablets:
[All texts read the same.]

THE MOVING FINGER WRITES
THE SLASHED WRIST BLEEDS

[Before men learned to write they lived forever
in caves and glades beside their human fires.
Hieroglyphics tombed the stiff Egyptians
Ashurbanipal was cuneiformed, the Greeks
impaled on Aristotle, Rome declined
beneath the weight of Gibbon.
History is their epitaph, this poem mine.]

THE FINGER WRITES
UNTIL THE BLOOD RUNS OUT
SUBMERGING
IN THE UNFATHOMABLE EARTH
MERGING
INTO THE WINE-DARK PAST

Samothrace

This man I lived with
back in graduate school
had a replica of the Winged Victory
submerged in the toilet tank.
He stated
it kept him regular.

And sure enough
I get this funny feeling
whenever I lecture
on the uses of the past.

The Owl Pellet

At tree level owl and professor blink
yellow noon, doze in the musty
hollow of tree and office, ruffle
dreaming of things furry astir by dark.
Below on Founders' steps two boys
dissect the pellet from the owl's late hunt,
catalogue the indigestible
debris of bone, claw, fur, one perfect skull
its jaw askew, recognizably rat.

Young friends, you are on the track: classify,
enumerate, set down in your tablets
THE OWL HAS MADE A POEM, THE GREY PROFESSOR
HAS VOMITED HER HUNT. I will analyze
my latest for you: this image, students,
is carved from Gloria Spoletti's thighbone
unforgettable for twenty years;
here juts the profile of a blind black boy
seen from a passing streetcar, there the hump
of my old crippled fencing-master, rotten
with all unanswered letters. When the greedy
guzzle of living sates us and the bones
stick in our craw—we cough up a poem.
It clears our throat if not our consciences.

So go, boys, and do you likewise.
Learn the wisdom of the owl professor:
FLY OPEN-GULLET INTO THE DARK,
BOLT DOWN WHATEVER SCURRIES.
Noontime's time enough to cull
the skeleton from the feast.

Poet in Residence

It's good sport
you sweet tooth boys
to pick the brains
of a grey poet.

Prise up the lid
from the hollow stump
but mind
you don't get stung
you flower children.

It takes bees
to make this honey.

News from the Villa Iovis

This printed circuit astrolabe
this paper sextant
will guide you beeline from the welter of boats
to the swarming funicular. Diagonal
the square circled with eyes, ascend the street
honeycombed with shops then walled with houses
narrowing isolate lifting to unmanned vineyards
hushed with the view of sea. When trees stunt
with the weight of sun and lizards flicker
out of the crumbs of esplanades you are there
where Tiberius was.

Sketching the fragmentary labyrinth
of cisterns, passageways, the mesh
of ambiguities (opus incertum)
stalking the shadows by dead reckoning
I stooped to a vaulted chamber, found myself
in the still center of a whirring hive
of bees, like some insane Homeric simile.

I cannot say the heaviness of air
or what slow waxen crystals clustering
embalmed the rotten brick like candle drippings
from a dark scented flame. So mesmerized
by the dull bruit of ages, yet unstung
I felt my honied cells burst slowly, drain
down the white vaulting of my bones
stood gaunt, an insect
caught in amber light.

And I alone have escaped to tell you:
BECOME ALIVE TO DEATH AND WORD THAT LIVENING
this poem my bee dance.

Carbon 14

Science calculates
all things that live between the cosmic rays
and earthbound nitrogen in equilibrium
take in and give off carbon radiation
by rate fifteen point three disintegrations
per gram per minute and at dying all
cease to take in give off irreplaced
their whole accumulation pulse by pulse
the steady running down runs down the ages
to a half-life past five thousand years

So scholars chronologue
neolithic campfires Mycenean graves
and that swift carbonized demise of
old Pompeii so radiant our bones
gavotte in this sweet biosphere ah love
a woman of a certain age in this
age of uncertainty why should I count
my middle age my half-life come and gone
when this dead pencil and this scratchpad glow
suffuse the landfills of ten thousand years
and these close carboned lines of mine to you
glimmer millennially as smouldered Troy
so radiant our loves
in this sweet biosphere

The Woman with a Tic

When I came forty
that three-headed dog love death and poetry
took me in its teeth and shook me
like one gaunt grey coiffured
woman in the first class compartment
whose every thirty seconds head
wrenched perfect circle round
three times and shuddered
home to neck. How still we all
passengers and stewardess
every thirty seconds did
not look.

                    I ask myself
if she understood what Einstein
and I have learned: stake down your mind,
drive it to the dead eye of god—
every thirty seconds galaxies
whirl round your center. I do reject
Copernicus' more elegant mathematics.
Not the sun swings me. I the sun.

Lady I have pinned you to my eye.
Every thirty seconds our stillness jolts
the great dog.

Icarus by Night

From the ascending jet
the cities recede
like a wilderness
of expanding constellations
like the heroic past.
If engines falter
what to fall back on
what underlies us
but universal darkness
pocked with fleeing stars?
We do most fear to fall
into no thing
but falling.

They have blown out
even the flaming sun
by which God candled
this egg shaped earth
saw in its molten yolk
a stir of feathers
set it warm to brood
nested in orbit.
From this dark egg
we all have hatched
we Icarus, at moth
to a doomed star
now free-fall
out of time.

Final Strophe

We learned a lot from each other, Prometheus and I.
He taught me that the rising sun was a poached egg
to break man's fast. But when the sun would rise
it was I rose with it, wheeling through the bright aether
to my breakfast.
The sun and I ate his poached liver
and it was I taught him to stomach it
to thrive on it.

The wilderness where we roosted, Prometheus and I
spread out beneath my flight like carrion fresh killed.
He'd plot my course, teach me geometry.
It was I that told him, Man, come down to earth, plunge beak
in the earth's guts,
don't speculate, eat! devour! ravin!
What was the fire for if not for burning?
Prometheus learned.

> Every dusk his fingers rooted
> deeper in the creviced rocks,
> every dawn his wild limbs fuited
> men to plow and tend their flocks.
> In the fields his green hair sprouted,
> cattle pastured on his beard,
> people drank where his blood spouted,
> nations fed on his sweetbread.

Civilization, I think he called it.
Anyway
there was carrion enough and to spare.
But I was uneasy: from where he lay
following the great planes with his eyes,
deaf by their thunder,
he never missed the songbirds.
From my crag I watched them fall, shriveled,
their blood clotted with gall.
I learned to let them moulder.

Spying the land's lay I saw the grey streaks
widening
in the merge and mesh of scab into scar,
the earth crusted over with dead cement
wilderness in strips lining the roads
fences to keep back
the starveling packs of wild dogs.
When the strongest leap the barbs hung with
the white antlers of stags
the autos swerve to crush them.

Where the raging cities flare
where the sterile speedways blare
sinking through the dying air
I share with men their tainted meat.
In the alley, by the ditch
boy or dog, girl or bitch—
men nor vultures question which:
what lives they kill, what dies we eat.

Was there another road, old friend?
What were those words you spoke
before your talons grew, before your beak:
the wisdom dripping from your wounded side?
Polis, res publica, city—
why did they fall?

Nesting together at myth's end
while the flames rise we croak
the secret that we knew but could not speak:
ALL PERISHES UNLESS THESE THREE ABIDE,
AIDOS, CARITAS, PITY.
So fall we all.

Over and Out

Inside the hogan
winds flatten into walls
the engines fall away
eight-sided space compounds us.
Our words float like balloons
we suck our food
from our squeeze-bodies.
Mission Control is dead.

Inside our skulls
blood-skinned tribes continually
hand over hand mount up
our spinal columns, emerge
into the flare-lit dark.
Behind our lids
sand-paintings rasp to color.

When suns cool
and the last comet gutters out
companion
we'll light the nipples of my breasts
inside the igloo
light my five taper fingers, brand
into our stretched skin
this log
this only poem.